A Framing Job

A Framing Job

poems by

Ricardo Means Ybarra

Red Hen Press

1997

A Framing Job

Copyright © 1997 by Ricardo Means Ybarra

All Rights Reserved

No part of this book may be used or reproduced in any manner whatever without permission except in the case of brief quotations embodied in critical articles and reviews.

Red Hen Press is a division of Valentine Publishing Group.

Acknowledgments: These poems appeared for the first time in these magazines and anthologies: *The Kenyon Review*; *Poet Lore*; *Blue Mesa Review*, *Café Solo*; *Electrum*; *OntheBus*; *Revista Mujeres*; *Bakunin*; *Poetry L.A.*; *The Northridge Review*; *Colorado North Review*; *Blue Window*; *L.A. Weekly*; *California State Poetry Quarterly*; *Grand Passion*; *The Poets of Los Angeles and Beyond*; *Cool Salsa*.

Cover art by Tom Trujillo
Photographs by Marlene Joyce Pearson
Book design by Mark E. Cull

First Edition
ISBN 1-888996-05-6
Library of Congress Catalog Card Number: 97-69605

Red Hen Press
Valentine Publishing Group
P.O. Box 902582
Palmdale, California 93590-2582

for Rafael

"Things fall apart
in our houses"
— Pablo Neruda

Contents

11	Whales
12	Smokey
16	Ants
17	As Birney Feared
18	Trout Fishing in Oaxaca
22	We All Knew
23	Trailer Park Cat #2
24	Corrido for Pete Wilson
25	Angels on the Highway
26	The G.E.
27	Samuel García
28	When she came to the door
29	Thursday
30	4:28 am
31	On a Ridge of the Santa Monica Mountains
32	Husk
34	Construction Has Been Very, Very Good to Me
36	Abreojos
38	The Bath
39	A Framing Job
40	Grout Lines
41	After the divorce
42	Milk Crates
43	#2
44	The Day Labor of Jesus Bautista
46	A Lost Trailer Park Cat
47	A Professional Woman Loves Her Body
48	Wake for a Sheet Metal Worker
49	The Only Child
50	February
52	Why I Never Got a Tattoo
54	Women
55	I Listen to My Son Grind His Teeth While Sleeping
56	Bear
59	Rain
60	Mixed Blessings
62	The Ants

63 Watsonville
64 Marilyn Monroe in Korea
67 Comet
68 Rafael, in the Garden

A Framing Job

Whales

The whales are here
again, on the other side of where I sleep.
I won't see them
I never see the whales
they don't fling themselves around for me
sucking it in for the splash
great grey flukes quietly disappearing.

They stay below
wait for me to crawl into my bunk
in this boat like a tunnel,
an attic, a glass of stale water
they wait so I can listen
breath of a whale
krill and kelp
the exhaled bubbles that stream against the hull
patterns of traffic lights.
And they rub
their skin, or is it eyelashes
to clear the long sad slit of an eye
to see reefs
or other boats.

I do what any man would
put an ear in the pillow
wind my watch
rub the hair on the inside of my thigh.

Smokey

> *Que bonito es querer como quiero yo*
> *Poco a poco me voy acercando*
> *a ti*
> *Poco a poco la distancia se va haciendo menos*
> "Llegando a Ti",
> — Javier Solis

Little by little I am getting closer to you
little by little
my name
my tag
the placka of spray cans,
is no longer restricted to East Los
confined by Freeways and the L.A. river
my name
is in your 'hood
on your garden wall.
Sure you'll paint it over
with fast rollers
but after the white paint dries
on stucco or block
it's still there
you know 'cause you feel me in those walls
under dry paint
Chuy, Smokey, and Flaco got down here.
Little by little, how beautiful it is to love like I love you.
Little by little the distance is getting less.

BIG TUJUNGA CANYON

Road Camp with my homeys—my family.
Chuy, Loco, Shorty, Kiki, Droopy, Danny, Pokey and
Jesse the chingon artiste, the needle man, a veterano.
Bare chested I look around at them
but I don't see your barrio

I see Sheriffs
the USC Medical Center
that's why I want the lady on my back.
I'm not talking about the monkey woman
Frida pinche Kahlo on a Levi jacket
or Linda Ronstadt singing Cielito Lindo,
get it right, canned peaches
stoned on home brew
watchale, this ain't my J.C. Penney tee
this is the real thing
my skin
ball point pen, needle and ink, do it Jesse, poke me full of holes
give me the lady, the Virgen, La Madre on my back
gonna protect me from the chingazos
of the man, the chotas, the fuckin' Sheriff.

The Sheriffs

Nighttime is crocodile time
sweaty snout
steady eye of searchlight
and the blossoms of red
cherry lifesavers that don't melt into the night.
Always two white boys from Altadena
sour with coffee stomach at 7-11, Winchells.
Is it the bird that cleans their teeth
the dash mounted computer clack,
or is it breaking twigs, the hunt in the swamp, Ramona Gardens
that tightens brown shirts around arm muscles?
Nightime, papa crocs got their tails swinging.

La Novia

"It was just a birthday party," she told the reporter.
"There were no problems 'til the chotas came.
"He didn't do nothing. He was trying to calm them down."

13

Lanterns, a string of them across the porch
plugged in Japanese lanterns
thumb tacks to hold them in place.
When one swung loose
she could feel it drop
that's what made her look
made her turn from his shoulder
where he had attached her
three days out of road camp
three days of love
poco a poco, they were getting closer
but the swinging lantern
the party had to be right
only three days when she pulled away from him
felt the electric charge off his skin
not wanting him to help
on the porch
the lantern in her hand
thumbtack between fingers
pressed into skin when they came
when she screamed his name
the lantern glow in her hand
only three days of love.

JAVIER SOLIS

I started out singing songs about women and broken hearts
because that was the only thing to sing about.
Now, I have fifty charo suits with silver buttons
a cigarillo while I sing
a tequila or two between songs.
They say I will die because of women
some sancho will find me with his wife
use a gun or a filero.
I would prefer the gun
less to think about

if the cabrón can shoot straight
then, of course he'll shoot her too
that's why they come to hear me sing
songs about women and broken hearts

La Madre

"All people take care of their children, but now you must protect them from the police."

Smokey

My name is Arturo Jimenez, a.k.a. Smokey
You don't know me, do you?
I was shot in the chest at a birthday party
three days after the Rodney King beating
three days out of road camp
three days back with my girl.
They left me on the sidewalk
wouldn't let the ambulance through
I don't know but I think I was already dead
'cause all I remember is the Virgen
trying to calm the Sheriff
trying to hold back his finger.
Nineteen years old.
Poco a poco, I'm getting closer to you.

Ants

As I shower this morning
I hear their songs of love
for one woman, the many children,
soft dirt, the new addition to the nursery.
Maybe they wait for the sound of pipes
the clang of hot water
then rush from the foundation
to cart away my things.

Yesterday, it was a toothbrush.

As Birney Feared

A dream has frightened me,
Birney wouldn't talk.
I want to hold him close
till his beard tears my cheeks
and tell him I love him
that he will have a family one day
it's no big deal
you don't plan them
they're like rogue tomatoes that surprise you in the mulch
among the ornamental plants.

In that dream it's as Birney feared
all the nails in the attic have been pulled
rafters, beams and shingles are floating
the flues and vent pipes hang
it could all come sliding down
or it could just flatten
the roof beam in the middle
shingles flapping like venetian blinds in open windows.

It worries me,
he's anxious about the day coming, the wind
he notices that the insulation has been rolled up
and thousands of sea turtles are hatching, scrabbling
between the rafters.

Trout Fishing in Oaxaca

What I saw in the Culver City alley
at 6:30 am
was a man
before the police arrived
sweater bunched up the middle of his back
no socks, one shoe missing
the weather that kind of L.A. drizzle
as the bus pulled up on Venice Boulevard
but didn't stop
and I knew I was alone
before mothers
gently push their children to school
button them close and hug them
whether they want it
knowing it's drizzly and will lift
like termites
ecstatic after the rain
free to beat tender wings for ten minutes
celebrate wicked pleasure
before the wet earth calls them back
that's how I knew
dried blood black
thick enough to scrape up.

Fixed in a tired Ford truck
needing a wash and a wax
I wait for Angel to open the door.
Together with coffee spilling out of the protective lids
we will admire green lights that speed us to our job
hope for the Gordo
lotto numbers
I always play the same
and tell my friend if I win
we'll build a small hotel in Oaxaca.

I wonder what he's thinking
behind a locked door in this neighborhood
undocumented
and what pension with me or health
I can't afford mine
yet we've opened and closed the days for eight years
so I know his family
know he sends la mayoria of his pay to them
and when I visit they treat me right
of course, that's the way
juntos in the truck
work is not the big thing
whatever it is has to be done
we mix our hands in cement
we swear when tile goes wrong
we cry out, somos artistas

and spend too much time
with women in their showers and baths.
But now a man nudges me
watches my every move
touches the careful layout
in intimate places.

I want to honk the horn
but the police have arrived
children held back
before Angel opens the door.

Two days later in Oaxaca
not the city
his pueblo after three hours of climbing
Macuiltianguis

another truck, the morning again
home brew mezcal
a twisted spine spills out of the jug
the color of coffee.
In the Sierra Madre there are no green lights
the bus doesn't stop on Venice Boulevard
the chickens, pigs, and goats bleating
before the blade

before the beginning of Lent
if I can lean in the heat
against the raw adobe of his family's house
no dead will find me in alleys.

The night
I finger the chainsaw blade
hanging with ancestors
on the wall of saints
my blood tastes cool
with tamales
and the start of Carnaval

a rocket explodes
revolving light over the rooftop
sirens
the children rush outside.
I could lock the door
cover my eyes with the burnt corn husks
and still see their faces
but Angel won't leave me behind
sure of the layout in familiar places
the firework towers in the plaza.
A wheel is lit
one missing shoe spins

in an alley
I suck the bloody finger
a horn dark at the tip.

Finally, it is time to fish
in a stream after a descent so difficult
I couldn't find the open sky
then I baited the hook.
Angel caught trout
beautiful tiny trout
maybe six inches the largest
exquisite, gasping gems
warped backs a good handle
gills blinking furiously
as if I could
rescue the tender red wings
calm the tail shudder
a foot
slapping the wet rocks.

What did I see?
I caught nothing
trout smell hesitation
strong as coffee spilled in a drizzle.

We slit them and slid them on stripped pine branches
I knew mine was cooked when the edges curled up, twisted black.

We All Knew

I knew that if the wind blew
one more day she would leave
she was wearing her sweater and long pants
even though it was August
near the middle
and she brushed her teeth
long and perfect

for my part
I walked around nude
just the feel
of the sun against my thighs
made me feel better
I went outside
to spray down the house
to watch the steam rise
the windows melt
into the healthy oleanders
the bougainvilleas around the back door
neighbors stopped by with cranberry juice
bananas, figs, zucchini
and she bathed every few hours

I tried to sleep
in bed her brush driving through thick hair
the curves
unnerved me
I moved to the living room
where pictures curled
sand against glass

sometime in the early morning
the wind slipped away
I opened a few windows
turned off the lights
the figs and bananas were heaped on the table
the zucchini had sprouted
flowers the color of headlights on a damp morning.

Trailer Park Cat #2

Finally the tile let go
a yawn
one piece clattering into the porcelain tub
I waited
the water heater fired up
refrigerator hum
those men on the roof again
stepping on persimmons, spitting
kicking at dried cat shit
loosening shingles.

Perhaps it's the photos that caused it
28 Polaroids
heaped on the floor
under a wall of stained borders
Max 1, 2, and 3, Shadow, Abby, Toes, Stella, Jamaica, Patches
in unposed positions with red eyes.
I hoped you would stay.

I heard another tile fall
remembered the candle burning in the pumpkin on the front porch
I was worried but couldn't leave the Polaroids
for another face
in a pool of white wax.

Corrido for Pete Wilson

When Dick Acuña sprayed his armpits with 25 cent cologne in the
bathroom of The Office Bar he had to stand on the tips of his work
boots
Sears Best honey smooth leather
sliced across the top from barbed wire
he had to stand so he could mimic a ballerina
in dirty white t-shirt, eyeglasses and toes
of sand running out the splits
California desert sand
in the stitches and on the concrete floor
under the urinal and the colored and ribbed prophylactic machine.
Orale, orale viejito—he yells
Dick Acuña is the funniest man I ever knew
he calls me a little old man, maybe
because I can hear my father gunning the '49 Ford pick-up painted
white with spray cans from Pep Boys or because I can feel him taste
the warm Budweiser and drink it anyway quickly
but I can't stop laughing as Dick Acuña dances
wiggles his hips, arches his back, spins for each white t-shirted
armpit, the Old Spice so sharp I looked in the mirror to see
who had slapped my face.

I had no fear of sand
sure I heard footprints
but I could barely speak la lengua Materna so how would I know
the secrets told at my bisabuelas house in Pomona on Sundays.

I didn't know from any of that
la migra, coyotes, pollos, mojados,
paddys.
I was with Dick Acuña in the desert
I pushed the 25 cent plunger with both hands.

Angels on the Highway

More and more I am followed by hair combs
blond wings
the constant rearrangement of halos
a divine countenance
shiksas in Rabbit convertibles.
Why have I been chosen?
A simple sheepherder
with the name of Jacob
I didn't steal my birthright on the PCH
wrestling with pink fingernails in my rear-view mirror
furious punching of radio
more Bad Company or Led Zeppelin.
I would feel better about this vision
if there were a Stairway to Heaven, or
just a ladder.
Nervously, I bless myself
spectacles, testicles, wallet, watch.
Angels,
blessed inches from my rear bumper
Gabriel in their back seat.
Let us pray for the intervention of disc brakes.

The G.E.

I run my tongue over teeth
and find a new one.
In the refrigerator the light
has burned out, I am worried
where to put my hands.
It has been like this for two mornings.

In front of the window
a lady walks her dog
an immense dog
he carefully sits on the concrete patio
staring, troubled
losing his hair.
I touch my tooth
and leave for work.

When I return
I see that someone has eaten
my bag of peanuts
a few shells bob in the carpet
still salty, full
and there is a note on the refrigerator
from Sid's Appliance Repair,
      come back darling
      all is forgiven.

Samuel García

His hand knew machete
but it held claw hammer

Maybe he was thinking about chocolate
for his daughters, and the

Twenty dollars he would send them
twenty dólares, twenty dolores, today

He came early
for two daughters sleeping

He would bust his ass
break his neck

Two wrists and collarbone
twenty dolores

His fingers clawed at the wet roof
as if he were pulling their covers down

Hand over hand
full of blanket

Slowly, so as not to frighten them.

When she came to the door

hammers stopped and they set to
pulling bent nails
the foreman stayed in the rafters
joints newly set and bolted
the flow of wood
his fingers white on the grain.

It was the sunset of nuns
walking to rosary
a man looks at an old letter
a woman brushes her hair on the porch
a wind sets on wet plaster
and
the blood hot whine of a 16 penny nail
pulling away.

Thursday

It is enough that I open my mouth.
I have found the morning
it climbs down the rain gutters
and slouches against my windows
like a reflection on white stone
termites, quiet in the foundation
cleaning wood
ready for the sound of latches.

4:28 am

That woman is here
with her wool, wheel and knife
she sleeps as the soft face of the clock
spins onto the floor.
I don't rest
in that content orange light
I count my mistakes
and a buck twenty-nine on the dresser
a flat bill, laid out.

She comes because I call for her
then I follow, careful
to act busy in the shadow of the laundry room
busy with calls, letters
as she goes on in her usual way—
cooks spaghetti, sticks her hair to the walls of the shower.
I have never asked her to leave
and now as her breath
presses against a pillow
I can't sleep, I listen
for the dog across the highway
to throw out his fear of cars
for the lights
to pierce his wire fence
and find me here like this.

She'll leave when the two
sharp arms of the clock
close on still threads.
A woman knows, after failure you go on.

On a Ridge of the
Santa Monica Mountains

Sunset hangs around like the dog
waiting for me to throw his Frisbee
slobbering, chewed on the edges
deep red.
If I throw it
a good toss down the hillside into the manzanita
loaded with rattlers
the dog might not return.

The ranch, a house and the just filled pool is far from the city
light, my trailer in Malibu
lamps over dinner tables
the front door porch
a tv
people coming home.

I don't know what is missing
my tools are in their buckets
the 120 foot long pool I tiled
with 3 inch green tiles
a lap lane
and seahorse on the first step
reflects a lingering vapor trail

if I would just walk to my truck
the lights will lead me down the mountain.

Three bats swoop past
another mosquito draws blood
a toad complains of sore shoulders.
What have I left
at sunset
drop my dirty pants
slip into dark water.

Husk

It was only 7:30
and already I started to think of the rest of the day

I dragged my surfboard up the beach to dry sand
sand that was still cold
under a fierce wind
out of the Northeast
cold as the rubber of the wetsuit
my feet and hands swollen
pumping blood
I peeled down anyway
opened my back and chest.

In the big waves of an Oxnard beachbreak
I was worthless
a fin loose, the surfboard must be repaired.

It broke on my only good wave
a thick mother
with the stones
to split me wide
the wave sucking up lost fishhooks and engagement rings
as if it knew I was hopeful
lost a step
one more prayer
bent knee
one more good wave
stiff bones and back
if I could hang on to that first drop into the pit
if I could just make one more leap.

It began with the cracks
I hadn't fixed
water had seeped in
weakened the fiberglass of the surfboard
brown streaks, old husk
and as the wave tore the fin
I knew that soon I wouldn't make this trip
saw the leak around the sink
ants burrowing through the grout
streaks of rust
termites tearing into our front deck
exposing the marrow
another crack
one last beautiful wave
water finds its way.

I could say I had all I wanted from that one wave
the spread of dawn out in the surf
my hair matted, thick with salt.
I rubbed my hands
to free the yellow grains
sand defiantly attached
in the web and under the nails
joints stiff, fingers wrinkled, trembling.

Construction Has Been Very, Very Good to Me

The lonchera, roach coach, pulls up
tootling La Cucaracha.
José, the driver knows the joke
and knows that in the cold rain of February
we are glad to accept the comfort of raised awnings
weak coffee that never cools in styrofoam
tacos at 8:30
ice fresh and crunchy as the new day around the Coke cans
bottles of Tehuacan and Jarritos. It could be a T.V. spot
except that José doesn't dance for us like Michael Jackson
it's enough that he remembers our names and offers the whining
tight-ass contractor the biggest green chile from the bottom of
the can.

When José leaves
the only shelter is in the back of the house
architectural dream
the Westside not the Valley
a remodel, master bath and bedroom
new pool faced with black granite.
Those that speak English only
take over a corner of the lumber pile
the majority sit on a stack of drywall or wherever they can find a
dry spot of concrete
the wind off the brackish water in the swimming pool
smells but we dig in anyway
a piece of chicken is dropped
the Salvadoreño swears at it, dusts it off and tells us
that now the pollo tiene sabor
the gabachos kick about gays in the military
someone quotes Howard Stern.
Lunch is more of the same only fifteen minutes longer.

That afternoon my helper and I finish the shower and steam
bath. The baboso also known as the contractor won't pay until
I've thrown in about $300 bucks worth of freebies
it's the owner, he says, '93 is going to be a tough year for the
wealthy and softens the blow by telling me a Hillary Clinton
joke.

What about the beauty of our work? The craftsmanship? The
bonding of workers? A six-pack on Friday afternoon?
Let me tell you, if you ever get a call from Pat Keegan,
contractor
take my advice and tell him to kiss your ass twice
then dump the rest of your cement, thin-set, grout and tile
cuttings into his new pool.

Abreojos

The morning after the storm
I saw the rocks through the steam of my first cup.

The morning after the storm
Abreojos
against the new light.

It was the surge and the motor that pushed us
between the rocks like a stutter
at the entrance to the bay
after the night of wind and waiting in the trenches
the bow pointed at the waves
hands at first holy communion
kelp, spread out from the rocks
lay against the boat hull
as if to slow us
the attraction of the propeller
a lover, naked with long brown hair.
Annoyed, gulls and pelicans
flapped over us
as the boat cut the dark strands
of the morning on the doors
the pink and blue and turquoise houses
Abreojos.

All hands were tight around cups of hot coffee
the four of us, topside
tired of the storm and the sea and
this boat rigging
clanking like dentures in a dry glass.
No one spoke.
Perhaps the relief had emptied our voices
our tongues, eyelids in the dark
whispered a plan to each shut ear.

In the empty bay
the crisp and silent light
I want to tell them—Abreojos—open your eyes, watch out
on the morning after the storm
houses of pink and blue and turquoise
heaps of cans bent as old shoes
white doors
Abreojos against the hill
of crosses leaning among rocks
dirt, a shirt buttoned to the throat
mouths open to the sea
dried shells of abalone on the beach.

We drifted in the spread wings of kelp
looking for safe anchorage
until we saw the man on the hill
or was it a woman
with legs wide, a red baseball cap
throw down the pick and shovel
behind a truck the color of dry mud
behind the broken tailgate
on the hill that bends the pick handle
bunches the cords of the arms
alone, dark hair spilling from the cap
so there is no mistaking it now
as coffee grounds settle in teeth
and the silence of cans turns brown in piles
thrown from the doors
the pink and blue and turquoise houses
Abreojos.

The Bath

In the tub we play games
with the tugboat and the wind-up hippo.
The water bounces in our close sea
sliding onto the floor
soaking the clothes thrown about.

"Mihito, wash my back," I say
and he clambers over my legs
exchanging places
forgetting the soap
as he scrubs so generously
tickling me with the loofa, laughing
"Wash daddy's back,"

then quietly, the water finally still
with no hands he pees on me.

A Framing Job

It was local work
scattered cigarettes in the tool box, the sports page,
long breaks
and then the frogs came.

I couldn't understand why they were there
attracted to the wet wood
green lumber that spits tree juice.
And, the frogs could climb
as fast as I nailed
they settled over the shiny heads
thin membranes pounding in unison
on wet dripping wood
sometimes touching the metal still hot
from penetration,
and with their black eyes, like coffee
women over cups
who talk easy, warm
staring me in the face.

At first I laughed and threw them on the other men
then the hammers fell.

Grout Lines

The ideal is straight
level and plumb
but not on the Harbor Freeway
Thursday morning.
I pass a fire truck
one firefighter stares
I wasn't expecting a straight lane
the emptied skyscrapers like shower stalls
steam rising from South Central
the Harbor Freeway a drain clogged with ash
pigeons roosting in Dodger Stadium
buses emptied of old graffiti.
I know smoke stains white grout lines.

I've been handcuffed, esposas we call them.
Maybe it was the goatee
did I look bad
a cholo
or a brother
but they put me down
flat down
my tile saw in the back of the truck
guns and assault rifles locked and loaded
black boots in my face
I wasn't wanting an ass kicking
black boots in my face
smoke from a thousand fires
tell me his name
while I kiss the ground.

After the divorce

small candles make Sunday difficult
and parking at the church where
my grandmother forces me to look
into this place
dark as a drive up the coast
when night has carried off
the beach swings and Jungle Gyms
left ice cream and towels
in the middle of the road
packs of dogs waiting
under the guard towers.

I know my grandmother's with St. Teresa
in the alcove
of the 25¢ candles.
Beautiful St. Teresa, the nuns said she used
crucifixion equipment to stop the ache
a martyr
waiting for me
to light another
of the thin matches in the center
of my palm.
But I won't do it.
I lit them all when I was a kid
until the light climbed into
the hard wood of the pews
and I'd lay there
stare at the smoke
and think of her.

Milk Crates

All week he throws plaster
whatever the cost
tape around the groin of Magic Johnson
a bullet train leaves Tokyo
a sheet curls into the typewriter

what they want is flat and blank
high density, low level
and the old man on a pyramid of crates
white wings
level, level
this is a modern house

what he needs is wind
not the air filtration system
the El Salvadorean maid smiling
wind to dry the plaster

he hears a child
wonders where the six toed cat is
the magnets on the refrigerator
finger paintings and Popsicle sticks
only 3 pieces of furniture

3 pieces of furniture
Euro-style, a flat black telephone
and as he rolls up his tarp
wind, pollen from the lime tree

he didn't dream of wind
that damn pick-pocket
yet there it is
a blunt fingered gypsy boy
to open the door of his Chevy pick-up.

#2

At daybreak I go to the mirror
no need for light
I just touch the switch anyway and brush my hair.
Should I shave?
I don't know, maybe if I could talk to you
about this loneliness.

Rainwater fills a metal cup on the deck outside the room
rust grows on the thick grains of redwood
I see your hair spread out on the bed
with no pillow, the sheets pulled from
the corners, breasts like sleeping
kittens, eyes closed.

I stare at my dark outline
feel lips, the rough texture of my face
I know I should shave.
Admit it, the morning confuses you.

El Day Labor De Jesus Bautisa

El sonido de escaleras siguiendote
Manzanillo
esa piedra escondida en su zapato
Ozumba, Nepantla
el coyote detrás de las rosas marchitas
Tepetlixpa, Popo
y la rica
ondulado sobre tu colchon de mangos de pala y de yeso mojado.

Cuando corto un limón
veo tu lengua
presionando a la gran cuesta de automóviles bonitos
de gringas en convertibles de BMW
cuervos danzando en Beverly Hills
ese vídrio moqueado
aliento de cabros de tu quinceañera hermana.

Manejo muy lento cuando llego a las esquinas
de huesos en gorras de beisbol
voces
gargantas de alambres partidos
manos con las palmas hacia arriba
el cemento de ayer.
Jesús Bautista
no puedes parar y esculcar tus zapatos
tu sabes que significa un six-pack de Coke sobre mi tablero de instrumentos
la caja vacia de mi camioneta
aquellos listos para poner su piel a las palas,
la cara redonda para la mierda de gordos perros americanos
anis el los arroyos
otra vez la piedra para un día de trabajo
mirando gorriones que bajan a las albercas.

The Day Labor Of Jesus Bautista

The sound of ladders follows you
Manzanillo
that stone hidden in your shoe
Ozumba, Nepantla
the coyote behind stale roses
Tepetlixpa, Popo
and la rica
curled on your mattress of shovel handles and wet plaster.

When I cut a lime
I see your tongue
pressed at the long beautiful automobile sunset
of gringas in BMW convertibles
dancing crows in Beverly Hills
that snotty glass
goats breath of your sister's quinceañera.

I drive slow when I reach the corners
of huesos in baseball caps
voices
throats of split wire
upturned hands
yesterday's cement.
Jesus Bautista
you can't stop to search your shoes
you know what signifies a six-pack of Coke on my dashboard
the empty bed of the truck
those others ready for skin on the palas
the curved face for the shit of fat American dogs
anise in the canyons
the stone again for one day's work
watching sparrows fall into swimming pools.

A Lost Trailer Park Cat

I am afraid of this level spot
near the ravine where wind stops
where coyotes watch flat shadows.

Even my cats
always somewhere close
wait for the open back door.

It does no good
to fiddle with the lights
to arrange the photos
the moon will continue
under our Eucalyptus
weaving a nest I cannot find
leaves will blow a scuttering rake
the neighborhood mutter
why doesn't he take care of this?

I know I must gather up the frightened cat
the one with eyes that follow the back door light
ugly colored fur
wattles that drag under her stomach
like a breeze through dry weeds

and carry her to the ravine
before the wind starts
to unravel our place.

A Professional Woman Loves Her Body

She folds long fingers carelessly
over a flat belly
feels for softness
loosens her dress to touch hips, thighs
on her 36 year old body.
Poppies drop petals on the table
the dryer begins another cycle
the apartment is not lonely.

That night after sex
they eat a watermelon in bed
crisp red meat
she remembers a biology class in Jr. high—
all females carry from birth
a limited number,
she cleans up the melon seeds
wraps them in newspaper
and sends her lover home.

She calls her mother
tells her she will move to France
quit the firm, the profitable career
she wants to hear again
about her habits as a child
a little girl.
Later, she waters plants
in the un-glazed pots
slips in three seeds, and
worries about their dryness
the urgent rush of earth.

Wake for a Sheet Metal Worker

It's their hands I notice first
thick-scarred, cracked fingernails, new cuts
swollen from working tin.
Sure hands that I hold too long
against my own, soft now
in their palms.

They're all here
sheet metal workers
and the sons of sheet metal workers
guys I grew up with
union men in good suits, good money
solid women and their kids
running and yelling
in the backyard
where we once ran wild.

We drink beer out of the can
in the driveway
no clouds on this twilight afternoon
wind lifts the corners of the street
fingers freeze
comfortable around thin metal.

He was family
a father's friend
Corky.
Tin knockers
they taught us their trade
and now my father hangs close
his wide fingers
pound flat on my back
tells me, four or five times
how glad he is that I came, as if
I could have forgotten
the square edge
a sharp clean cut.

The Only Child

It starts when your son
feet cold at sunrise
comes and curls tight in your bed
thrusting toes like stream pebbles
where the hair stops, almost to the groin.
You watch him slip off into sleep again
to run through fields
past lizards that flex in early sun.
You can't stay in bed
his feet like embers.

You think of LA
Silver Ridge Avenue, Echo Park
crowded with streetcar rails
weeds and flattened pennies
hobos who can't sleep
your metal lunch box clanging
the nuns with small white faces.

And you remember your grandmother
in a sanitarium
still grieving for her lost child,
where lights dimmed daily
in electric air
hands folded, blue claws
shaved circles that brightened her temples
the silence
as she gazed out on the hill, the still fields.

It starts when your son runs hidden in tall grass.

February

Today, after I have left
he will come around
to pry off the door knobs
as if they were carnations,
his slippers, the wind from a bagpipe
or a bottle of gin.

He told me once about swinging at night
on a hemp rope, knotted below the seat
under a tree like a net
with a million open windows.
The stars were comets, he said
if you put your head back
and your arms ached
and if your legs were straight out
mouth closed
and sometimes, when your eyes shut.

I'll have to come back tonight, late
because I know where to find them
the knobs are in the old room used by us kids
where he painted his night sky
that once glowed before we slept
years ago when the net was open, and
he would burst through the door
his bald head
splitting the place
a shine of popcorn and gin, the laughter
falling around him.

Then, I'll lay back slowly on the bed
to look for those stars
so hard to see
when my dark is loaded down,
but I know he's waiting
for my eyes to shut
legs straighten out
arms to ache
and swing man
swing.

Why I Never Got a Tattoo

Beto was a bodybuilder
cop
ex-Marine
my father's best friend.
At parties he'd wrestle me
it never mattered how much beer
or if he had a new girlfriend
Beto would strip to the waist
hand over his pistol and wallet bulging with badge
to his latest sweet thing
get down on the floor
wrestler's position
not a hair on his chest
thin line of sweat
lips rubbering with each wind suck
flared nose
as I'd slap into place
hand grip the mossy road-kill
once an eagle
now a blue jay sucked under an Oldsmobile
over the Semper Fi
I could barely read
on an easy five pounds of bicep.
"Are you ready,
did you eat your Wheaties,
can you take a good ass thumping?
Then, call it out."

The Chumash
lovers of tattoo
plentiful abalone, lobster, fish and mussels
the women digging up plants, blanching acorns
no one to fight
and you can only fuck so much
why not tattoo
at Malibu
pierce the penis with sea urchin spines
why not paddle a dugout to the Channel Islands.

I've been across that stretch a few times
glad I had a motor
when water ran black
in the flex of waves
faint outline of land
a needle stuck under my skin.

I should have done a sweat lodge after one trip
or gone straight to Hollywood
taken my tattoo
on the neck or inner thigh
but I drank a six pack
curled into the pillow
set the clock for work.

Women

Why do you ask me about sex
say, "you must get it so much."

You think other women find me beautiful, that
I am caressed at stop signs
in shoe stores.

You long to hear that
I made one at the Beverly Hills Hotel
a young actress in Venice
wives, waitresses.

But do you know
when I am stripped
I lie in water
and weigh more than a 1000 lbs.
I cannot tilt my head
I cannot touch
they enter my flesh
suck on my breasts
lay over my body
exude a wooly lanugo

floating
blondes cover my eyes.

I Listen to My Son Grind His Teeth While Sleeping

Small stones
crumble as you speak
air leaves
sand on your lips
and you can't feel your breath
because of the toads
at the river
with nothing better to do
than talk of your hands
in the water
your hands growing moss
your face in your hands
your mother calling out
her room filled with tadpoles
as your hands hold stones
in the cold river
and your mouth fills with song.

Bear

Our hot tub was broken
and after six months of staring at the black hole in the deck
a four foot circular redwood Zinfandel vat
small by industry standards
occasionally shared
if it's hot enough
I called the Minneapolis Zoo.

I could have fixed it
an electrical problem
color-coded wires, the breaker off
explicit directions
but lately geometry interests me
measurement, properties
the relationship of points, lines, angles, surfaces and solids.
I don't like electricity.
I needed a bear.

The Minneapolis Zoo assured me
they wouldn't send me the bear
bear swimming circles
bear always swimming to the right
concentric circles
worrying the experts
elliptic circles
annoying the public.
They want to cure the bear
I want to fix my hot tub.
A geometric progression.

I borrowed the bear
it wasn't as difficult as it sounds
dressed as Pablo Picasso
the famous beret and striped shirt
we flew to Los Angeles.

Now, I have the bear and a hole in my deck
moving water, restless pleasure
a cylinder.
In the Minneapolis Zoo the bear swam right to left
always in a circle, there's a tight converging point
in a hot tub
we can fix on it.

The bear in my hot tub
dressed as Pablo Picasso
drinks a beer with me.
He can repair the electrical problem
it seems that it's physics
not geometry
electrons, neutrons and protons
polar opposites
but once it's fixed I'll have to ask the bear
not to pee in the warm water
that might be difficult
it's a common response
I admit.

The bear likes Greek olives.
I was prepared
made a Price Club run
I'm also well stocked with French cigarettes
and Miller's High Life
cheaper by the case,
fortunately Picasso couldn't join us
three of us would be a tight squeeze
none of us skinny
the hot tub not up to Industry standards.

The bear in striped shirt
smoking French cigarettes
beret cocked at a 68 degree angle
Greek olive pits spattered on the bell pepper plants
speak of Picasso's Blue Period
not Guernica
a series of pyramids
angled to the hypotenuse
jagged ice floes, frozen landscape
a bear crossing the broken expanse.

I'll talk to the zoo officials in Minneapolis
the hot tub is a hole in my deck
a bear with beret and striped shirt
bear who swims in circles.
It's not a principle of geometry
once the problem is fixed
the water will flow left to right.

Rain

The arrival of ants
and the wet weather
are noisy as the turning of cement.
But this is what I wanted
to lie in bed with you
to crawl among the bones
the smell of dark hair
blunt toe nails, sunlight
on tired wings
pine sap and lost teeth.

So what if I meet them
my face lopsided
the pillow case damp as a marsh
covered with cat prints
love letters
sheets the color of windows dripping.
Will the ants care that leaves
stain sidewalks
worms struggle to breath
or that my skin shines
because you are here
next to me
with your eyes of sliced watermelon.

Mixed Blessings

Chances are it won't ever rain.

On the one hand I pictured myself
stuck inside with Cynthia
under the covers
the drapes wide open
and while torrents of wind-driven rain whip the windows
we burst through cracked hardened mud after ten
months of slow breathing metabolism cranked down low

it's January for chrissakes
people crowd the beach
the cat wants to come inside
and then outside.
I'm edgy as an African toad and then again
Cynthia hasn't bothered to cover herself.

So why should I worry
the sun
a drunk who can't pick up his copper penny
continues to lean
in the afternoon
I see no reason to stop looking
even if I burn myself staring at her
like an idiot
it's one thing to turn away
and close the drapes
I'm bright enough to fixate
on beauty
since chances are
it won't ever rain.

I could pray for rain but it's impossible to keep my hands
away from you
an absolution in the glare off the windows
shower of holy light
my fingertips
as you cup the bed closer
the line I draw between your thighs
blessed on your bare bottom.
Chances are
the cat wants to come in again
a horn honks, asphalt cracks
the cilantro and basil
droop in the clay pots handmade in Tecate.

Even as I want rain
I love you
and while we witch for water together
what sun matters
but you.

The Ants

I found them in a hurry
crossing the kitchen sink

I should have understood the concern
they have for me
evident in their ragged lines

instead, I took my coffee to the table
dropped whole grain slices into the toaster
opened the paper to the weather report

it will rain today, and
ants helpless in rushing water

know I can't afford to miss
another day's work
still, they cling to my dirty dishes.

Watsonville

On the scenic central coast of California
stop in Santa Cruz and Carmel
not Watsonville
the gut of Monterey Bay
a yellow stink
after the harvest
weary as the urinal
of any Main Street bar.

In Minneapolis, Hartford, and McCall, Idaho
Mexican restaurants open
the new neon Corona beer sign plugged in
pots of Las Palmas enchilada sauce
simmering in the back room
tortillas heated by microwave
savage chiles
strung like firecrackers
warn the soft underbelly
a vision of señoritas
in white peasant blouses
tacos bent.

There are no redwoods gracing your house
only fields Watsonville
and after the harvest
what remains of you
a few bare stalks of brussel sprouts
rotting erect
discarded lettuce
stained and brittle as last weeks newspaper
waving good-bye
the last perfect strawberries
protected in plastic baskets
green as the work cards buttoned in your shirt pockets.

How will we know you were here
packed like hard red stones
on the final bus.

Marilyn Monroe in Korea

116th Medical Battalion

The smoke in the foreground
of two hundred cigarettes
sucked to flame
wool glove to blue lip
red eye of G.I.
gathers the tightly grouped heads, some empty trees
and the arrival of a flatbed truck
two days before Christmas.

I hold up another slide
want to know the name of those trees
as I squint, gather the hazy light, use the magnifier
drag in my breath with those waiting boys
at the sight of platinum hair, lips the color of her dress
the smile from the back of a flatbed truck
the American smile for each one of us
a whisper of dragonfly wing
on wet black rock.
Kodak transparency.

Su Yong Kim In Lynwood

His hands had been nervous for a long time
fingers and thumbnail stained yellow
on the right from the constant Winstons
he didn't light up for pleasure
or the hope of seeing Aldo Vega and
two other neighborhood boys enter his store.

Counter man, liquor store owner, lover of fermented cabbage
and family
a loaded pistol under the cash register
if I showed him the slides maybe he could tell me the name
of those trees

or if Marilyn
fatigue jacket opened in spite of the cold
sung a song or only waved, blew kisses
before she left us with her smile
a caress in frozen mud
and crushed cigarettes
a smoldering wing
the smile that brought Su Yong Kim to Lynwood.

Aldo Vega At The Boxing Club

Converted garage, wet clothes stink
kim chee or menudo
left, left, right
on the plywood floor squeaking under the mat
and these are the 12 and 13 year olds.
Aldo Vega at the Lynwood Boxing Club charges forward head down
traps his homey, sparring partner on the ropes
punishes the thirteen year old body
sting slap of leather
flinch
fists grooving kidneys.
Aldo doesn't worry when the counter-attack comes
to each side of his head
where it does no harm
let this fly boy pound
but Aldo never planned it this way
it happens so fast
the uppercut jacks his head
like the bullet on the street outside the liquor store
how'd he slip the uppercut
why'd he shoot him
the corner posts turning
his focus smoky
no sweet grunts.

Laying there he doesn't notice the poster of Oscar de la Hoya on
the back wall
holding out the keys to his new Corvette if you'll stay drug free
it's the bell he wants, the wail of ambulance
a far-off wave from a platinum blonde on the back of a flatbed, a
huera wiggling her good-sized ass in a tight red dress
smiling at him like she's sweet
you know what I mean
but it's cold and she's at a great distance.
If I showed Aldo Vega the slides maybe he could tell me
he felt her warmth, the eruptions off her milky white bosom
or is it just the blessed bell he wants
and her smile.

Cookies

"You waited for three hours to catch a sight of her and you can't remember?" I ask the Korean War vet.

At the liquor store in Lynwood there was a bag of cookies stolen
or paid for
a chase
perhaps a threatening screwdriver or knife
and Aldo Vega was shot by Su Yong Kim
over cookies.

Don't we remember
her promise
platinum blonde hair
lips as red as a pulled cigarette
the name of those trees
her breathy voice
fatigue jacket opened on the back of a flat bed truck
in spite of the cold
and her smile.

Comet

I finally found you
limp as a spider web
a termite wing
floating translucent
in the new heat of coming summer
out in the back
towards evening

a whisper
breath of one lung
the reflection on leaves
eucalyptus can't hide.

I wish that you wouldn't go
steady friend, know
there will be your
final wave
and then the constant sun
a dog barking
blonde fur shed on the carpet.

Without you
who will catch the long ball
hit to centerfield.

Rafael, In the Garden

The action is quick
his small hands press in careless dirt.
He doesn't talk.
The spade clicks
like small stones
in the habits of a stream.

He doesn't stop.
A leaf rests on his tongue
and he sees
the wide eyes
shallow prints on damp earth
a thin, black cat
the tree
his father in the shadow
his father gathering lemons.

Biographical Note

Ricardo Means Ybarra is an L.A. mestizo, a 6th generation Californiano born in Echo Park at the Queen of Angels Hospital, the year 1950. Raised in a labor union family, he was the first to graduate from a University, U.C. Santa Cruz, where he earned his B.A. in Latin American Studies.

Published in over twenty journals, he has received an award from the American Academy of Poets. His first novel, *The Pink Rosary*, was published by Latin American Literary Review Press in 1993. *Brotherhood of Dolphins*, his second novel, was published by Arte Público in September of 1997.

Ricardo Means Ybarra has dutifully returned to construction; a tilesetter, he lives in a trailer with his wife and rebellious sixteen year old near the beach so he can surf whenever he has the time and, hopefully, write.

Biographical Note

Ricardo Means Ybarra is an L.A. mestizo, a 6th generation Californiano born in Echo Park at the Queen of Angels Hospital, the year 1950. Raised in a labor union family, he was the first to graduate from a University, U.C. Santa Cruz, where he earned his B.A. in Latin American Studies.

Published in over twenty journals, he has received an award from the American Academy of Poets. His first novel, *The Pink Rosary*, was published by Latin American Literary Review Press in 1993. *Brotherhood of Dolphins*, his second novel, was published by Arte Público in September of 1997.

Ricardo Means Ybarra has dutifully returned to construction; a tilesetter, he lives in a trailer with his wife and rebellious sixteen year old near the beach so he can surf whenever he has the time and, hopefully, write.